Sop CEN

SANTA CRUZ CITY-COUNTY LIBRARY SYSTEM

0000111611604

D0601745

J 582.047 CHA
Chambers, Catherine.
Bark /

ACJ-4297

DISCARD

SANTA CRUZ PUBLIC LIBRARY
SANTA CRUZ, CALIFORNIA 95060

DEMCO

Would You Believe It!
Bark

CATHERINE CHAMBERS

RSVP
RAINTREE
STECK-VAUGHN
PUBLISHERS
The Steck-Vaughn Company

Austin, Texas

SANTA CRUZ PUBLIC LIBRARY
Santa Cruz, California

© **Copyright 1996, text, Steck-Vaughn Company**

All rights reserved. No part of this publication may be reproduced or utilized in any form or by any means, electronic, mechanical, including photocopying, recording or by any information storage and retrieval system, without permission in writing from the Publisher. Inquiries should be addressed to: Copyright Permissions, Steck-Vaughn Company, P.O. Box 26015, Austin, TX 78755

Published by Raintree Steck-Vaughn Publishers, an imprint of Steck-Vaughn Company

Library of Congress Cataloging-in-Publication Data
Chambers, Catherine.
    Bark / Catherine Chambers.
       p. cm. — (Would you believe it!)
    Includes index.
    ISBN 0-8172-4100-0
    1. Bark — Juvenile literature. 2. Bark — Utilization
— Juvenile literature.
    [1. Bark. 2. Trees.] I. Title. II. Series.
    QK648.C43   1996
    582'.047 — dc20   95-19538
                                       CIP
                                       AC

Printed in Hong Kong
Bound in the United States
1 2 3 4 5 6 7 8 9 0 LB 99 98 97 96 95

Contents

What Is Bark?

All living things need protection from the weather — the sun, the wind, and the rain. Animals are covered by skin and often by fur or feathers. Trees and bushes have bark to protect their trunks, branches, and roots.

There are many different types of bark. Barks can be rough, like this sycamore tree's (top). Others are smooth. The eucalyptus tree's bark is covered with patches (middle). And the paper tree's bark peels and flakes (bottom). But all of them do the important job of protecting the tree.

inner bark

outer bark

wood

Inside and outside

There are two main layers of bark. The inside layer is thin. It carries the plant's food from the leaves to the roots. The outer layer of bark protects the inside layer. It is both tough and waterproof. But it can bend a little, without breaking.

Amazing bark

Insects live inside of bark. Plants and fungi, such as mushrooms and toadstools, grow on it. Birds can find both food and shelter in bark. This woodpecker has pecked through some bark to make its nest.

Bark is very useful to people, too. In this book, you can read about all of the amazing things that people can do with bark.

7

Dress Up in Bark

Bark is made up of fibers. Some fibers are very long and stringy. People weave these into cloth. Fibers can also be twisted together to make rope or string.

From bark to bags

This man comes from Bangladesh, in Asia. He is carrying bark fiber. It comes from the jute plant. He will use it to make a thick floorcloth. Jute carpets are popular all around the world.

More than 150 years ago, people in Scotland found a way to spin jute into even thinner strands. The jute thread could then be woven into cloth by machine. The cloth was very strong. It was made into burlap sacks and even into horses' feed bags.

Rope from trees

This enormous tree is the baobab. It grows in parts of Africa and has many uses. The fibers of the inner bark are often made into rope.

Soft bark cloth

In the Amazon region of South America, people weave bark fibers into soft cloth. This Yuracare Indian woman is printing a pattern on some cloth that was made from bark fibers.

The Nootka Indians of North America used to make cloth from the bark of cedar trees. The cloth could be used to make warm, thick capes that were needed for the winter.

9

Boxes, Baskets, and Bags

Some barks can be bent without breaking. The North American Indians have used these barks for thousands of years to make boxes, baskets, and bags.

Stacking baskets
There are 30 baskets in this picture! North American Indians made them more than 250 years ago. They used white birch bark. This bark is very tough, and it keeps water out.

A bark box

Birch bark is very strong. But it can bend easily. The outer bark of the birch tree peels off in layers (see above). These layers can be sewn together to make containers. The birch-bark food box (left) was made by North American Indians. They used only a single piece of outer bark. The texture of the bark makes a very attractive pattern.

Flat cedar bags

This flat pouch is made from cedar bark. It also comes from North America. It is decorated with shells and dyed porcupine quills. You can also see glass beads. These were most likely bought from Russian traders.

Keeping Water Out

Bark protects trees from the rain. It keeps the water out. So some people use bark to make boats. Others build houses and roofs with it.

Keeping the rain out

Australian Aborigines made this bark roof to keep the rain out. The bark is cut from gum trees and then tied to poles. The house is built on stilts to keep the floor dry. The Aborigines also light fires underneath these houses. The smoke from the fire keeps mosquitoes from coming too close.

A boat with stitches

This bark canoe was made more than 100 years ago in South America. Some boats are still made like this today. First, the boat builder makes a long slit down a tree trunk. Then the bark is peeled off, in one piece. To make the canoe, he or she fits the bark around a wooden frame. The boat builder then folds up its ends and stitches them into place.

13

Bark That Bounces — Cork

Cork is amazingly useful. It is the bark of the cork oak tree. It is spongy and light. It is strong, and it even bounces! Cork can keep out the heat and the cold. It is covered with tiny holes, but it is still waterproof. Cork can float, too.

Cutting cork

Cork oak trees grow in Spain, Portugal, and North Africa. Cork is cut from the oak every ten years. A long, curved knife splits off large, curled slabs of cork. These are then boiled to make the cork easier to bend. When the cork is cut from the tree, the trunk underneath is painted bright red. Slowly, the bark grows back again.

The first cork that is cut is very rough. It is used for fishing floats, as a cheap floor covering, and for packaging. The next time the cork is cut, it is much smoother. This cork is used for bottle stoppers and smooth, tough floor tiles.

Bark and bees

In southern Portugal, people use cork to make beehives! The hives are made from large cylinders of cork. A circle of cork is nailed on the bottom to make a base. The thick bark makes a good home for the bees. It protects them from the heat and also from the cold.

On the shelf

Cork stoppers are often used for bottles containing oils, vinegars, or pickles. Wine bottles usually have cork bottle stoppers in them, too. The spongy cork squashes to fit into the neck of the bottle. It keeps air from getting inside. Air spoils wine. The cork is also waterproof, so it keeps wine from oozing out of the bottle.

15

Bark Books

People can write and paint on a flat piece of bark. Some barks can be made into a pulp for making paper.

Beautiful bark

This dazzling picture is being painted on amatl, which is a type of bark paper. It comes from Mexico. The Aztec Indians of Central America made bark paper hundreds of years ago.

Pages of bark

The whole book in this picture is made of bark. The bark was beaten flat and smooth to make the pages. It was made about 200 years ago in Sumatra, Indonesia. It is a book about magic. It tells people their fortunes from the stars.

Making paper

In Korea, paper is made by using the bark of the mulberry tree. An ancient method is followed. The shoots from the tree are boiled in water and ash. Then the bark is stripped off of the cooled shoots. It is boiled again and beaten into a soft pulp. In this picture, a man is flattening the pulp. Once the pulp is dry, it will form a sheet of paper.

Animal images

Australian Aborigines also paint pictures on bark. The pictures are often of sacred animals. This kangaroo has been painted onto a piece of bark that was cut from a gum tree.

Bark on a Plate!

Would you believe that you can eat bark? Bark fibers can be tough. But some barks produce gum that you can eat. Other barks are used to flavor food and drinks.

Making food shine

A sugary glaze makes this fruit pastry shiny. Some barks make sticky gums that can be eaten. They are used in glazes.

Nibbling bark

Goat willow trees, like this one, grow in Europe and in much of Asia. They have a very good name, because animals, such as goats, enjoy chewing the tree's smooth, gray bark! Goat willows are very tough. So they are not harmed by being nibbled.

Bitter bark

Would you like to drink bitter bark? Some people do. The bright orange color and tangy taste of this drink were made by adding Angostura bitters. These bitters come from the bark of the cusparia tree. They were first made in South America.

Spicy cinnamon

The strong, sweet-smelling cinnamon spice tastes good in cakes and in drinks. It is made from the dried bark of *cinnamomun* trees, which grow in parts of Asia. The spice can be bought either in the form of curled sticks or powder.

Bark Medicine

Would you believe that bark can make you feel better when you are sick? Some barks contain chemicals that are used to treat diseases. Others are crushed and spread onto wounds to help them heal more quickly. Bark medicine has cured many illnesses.

Treating malaria

When Europeans began to explore Africa and South America, they often died from the disease, malaria. But later explorers were luckier. An amazing cure was discovered — quinine. It is found in the bark of the cinchona tree (right). The medicine was first used by the South American Indians. But European travelers introduced it to the rest of the world. Now people know how to make the chemicals that are needed for quinine.

Witch hazel magic

Big, black-and-blue bruises on your knee can look terrible! But a cream made from witch hazel helps soothe them and makes them go away faster. Witch hazel comes from the bark of some shrubs, like this one.

Bark — the snail slayer!

These West African fishermen have to be careful! Water snails in the region carry a worm that carries a disease, called bilharzia. But the gray-green bark of the soapberry tree can be made into a poisonous soap. The frothy soap is put into streams to kill the snails and their worms, letting the men fish safely!

Sticky Bark

Some trees have sticky gum right underneath their bark. Rubber latex is a very useful gum. It is collected by slitting the bark.

Rubber tapping

This white rubber latex is being tapped from a tree of the spurge family in Malaysia. Every two days, a long, curved slit is made in the bark. Latex drips slowly from the cut into a cup. Once the tree has been slit all the way down, the bark is left alone to grow again.

Drying rubber

Here in Malaysia, a huge ball of rubber latex is being smoked near a fire to dry it. The ball gets bigger as a new layer of latex is poured over it. The latex dries into solid rubber and is then ready for people to use.

Rubber sandals

Rubber is tough but flexible — it can bend, stretch, and bounce. This makes it very useful. Car tires, rubber bands, and bouncing balls are just a few of the things that people can make from rubber.

Rubber can be recycled, too. This market stand in South America sells sandals that are made from old rubber car tires.

Leather Needs Bark

Some barks contain a chemical, called tannin. This is used to help keep leather from rotting, shrinking, or stretching too much. Today, however, factory-made chemicals are usually used instead.

A soft saddle
This soft leather saddle cover was made in Russia, 1,500 years ago. The best Russian leather is still treated with a sweet-smelling bark tannin from the myrtle tree.

In the tannery
Leather is made in a tannery. Here, a worker scrapes hairs from the leather before it is dipped into a vat of tannin. This picture is 400 years old, but some tanneries still use the same methods.

The mighty oak

A hundred years ago, oak bark was used in many tanneries. Today, it is only used to make the finest leather. Here, in the north of England, a woodsman strips oak bark off a trunk. He then farms the oaks, making sure that they grow back after wood has been cut from them.

Dipping the leather

These are tanning and dyeing vats in Morocco. Tanning takes several weeks. First, leather is dipped into a weak tannin mixture. Then it is soaked in stronger mixtures. Finally, the leather is left on the roof to dry!

Color It with Bark

Bark itself can be dull to look at. But it can be used to make bright colors for paints and dyes.

Bright red bark
A South American Indian is sewing this string pouch. The red patches on it were made from a bark dye.

Horse chestnut
This is the gray bark of the horse chestnut (conker) tree. It can be ground into a powder and made into a strong, black dye for coloring silk and cotton. The bark of the common alder tree can be used to make brown, black, red, green, and yellow dyes!

Bark ink

Bark can be used to make ink.
The lumps growing on this tree's
bark are called galls. They are
often started by insects. These
galls were made by wasps. Galls
contain gallic acid, which is a
kind of tannin. It can be used
with other tannins to make
different colored inks.

Keeping cloth bright

Bark tannin is also used as
a mordant. A mordant keeps
cloth dyes from being washed
out. It keeps colors bright.
Here you can see some rich,
blue cloth being dipped into
a dye vat. These dye vats are in
Kano, in northern Nigeria. The
deep, blue color comes from
the indigo plant. The
bark of the locust
bean tree is used
as a mordant.

27

Bark Crafts

Take a closer look at bark yourself with some arts and crafts activities. Here are some ideas to help you get started.

Bark rubbing

This is a great way to discover the patterns and textures of different barks. Simply put some thin, white paper against a tree trunk. Using a thick crayon, flat on its side, color across the paper, pressing it down on the trunk. Watch a bark pattern appear! You can use your bark rubbings to make pictures and collages.

A water snake

Make this water snake out of old bottle corks, string, and tape. The tape sticks the string to the corks, which keeps them together. You can decorate the snake with stickers, if you want. Put the snake in a bowl or a bathtub filled with water, and discover just how well cork can float.

What other uses can you think of for floating corks?

Cinnamon snacks

Cinnamon bark makes a delicious spice. You can use it to flavor all kinds of food. These ideas use powdered cinnamon:

- Try sprinkling some on top of a mug of hot chocolate.
- Mix a little cinnamon with some brown sugar, and put it on top of some yogurt.
- Cinnamon toast is good, too. Mix one teaspoon of cinnamon with two teaspoons of sugar. Sprinkle the mixture over some hot, buttered toast.
- Look through some cookbooks. See how many recipes you can find that use cinnamon. Maybe you can make some of them, too.

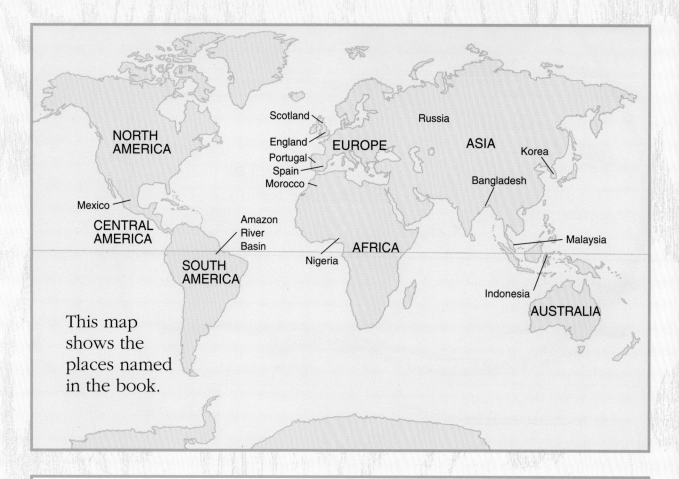

This map shows the places named in the book.

Further Reading

Diehn, Gwen and Terry Krautwurst. *Nature Crafts for Kids*. Sterling, 1992.

Greenaway, Theresa. *Woodland Trees*. Raintree Steck-Vaughn, 1991.

Markle, Sandra. *Outside and Inside Trees*. Macmillan, 1993.

Index

© 1995 Evans Brothers Limited

Acknowledgments

Editors: Rachel Cooke, Kathy DeVico

Design: Neil Sayer, Joyce Spicer

Production: Jenny Mulvanny, Scott Melcer

Photography: Michael Stannard (page 29)

Illustration: Kevin Jones Associates (page 7)

For permission to reproduce the following material, the author and publishers gratefully acknowledge the following:

Cover (top left) Gary Retherford, Bruce Coleman Limited, (bottom left) Werner Forman Archive, (right) Andrew D. R. Brown, Ecoscene, (logo insert, front and back) P. W. Rippon, The Hutchison Library **title page** (logo insert) P. W. Rippon, The Hutchison Library, (main picture) David W. Jones, Lakeland Life Picture Library **page 6** (top) Patrick Clement, Bruce Coleman Limited, (middle) Michael Cuthbert, Ecoscene, (bottom) D. Nicholls, Ecoscene **page 7** George McCarthy, Bruce Coleman Limited **page 8** The Hutchison Library **page 9** (top) Ecoscene/ Gryniewicz, (bottom) Tony Morrison, South American Pictures **page 10** The British Museum **page 11** (top) Louise Murray, Robert Harding Picture Library, (middle) The British Museum, (bottom) Werner Forman Archive/University Museum, University of Alaska **page 12** Tony Morrison, South American Pictures **page 13** V. C. Sievey, Eye Ubiquitous **page 14** Eric Schaffer, Ecoscene **page 15** (top) Medimage, Anthony King, (bottom) Robert Harding Picture Library **page 16** (top) Tony Morrison, South American Pictures, (bottom) Ronald Sheridan/ Ancient Art & Architecture Collection **page 17** (top) J. G. Fuller, The Hutchison Library, (bottom) Werner Forman Archive **page 18** (top) Tony Souter, The Hutchison Library, (bottom) Hans Reinhard, Bruce Coleman Limited **page 19** (left) Robert Harding Picture Library, (bottom right) Gary Retherford, Bruce Coleman Limited **page 20** (top) Mary Evans Picture Library, (bottom) Tony Morrison, South American Pictures **page 21** (top) Dr. R. Parks, Oxford Scientific Films, (bottom) C. Bryan, Robert Harding Picture Library **page 22** A. J. Deane, Bruce Coleman Limited **page 23** (top) Luiz Claudio Marigo, Bruce Coleman Limited, (bottom) Tony Morrison, South American Pictures **page 24** (top & bottom) Ronald Sheridan/Ancient Art & Architecture Collection **page 25** (top) David W. Jones, Lakeland Life Picture Library, (bottom) Ian Harwood, Ecoscene **page 26** (top) John Wright, The Hutchison Library, (bottom) R. Wanscheidt, Bruce Coleman Limited **page 27** (top) Jeff Foot Productions, Bruce Coleman Limited, (bottom) The Hutchison Library.